Energy Challenges And Opportunities

Philemon Chigeza

Illustrated by Kenny Estrella

Copyright © 2012 by Philemon Chigeza.
501390-CHIG
ISBN: Softcover   978-1-4691-6353-6

All rights reserved. No part of this book may be reproduced or transmitted in any form or by any means, electronic or mechanical, including photocopying, recording, or by any information storage and retrieval system, without permission in writing from the copyright owner.

To order additional copies of this book, contact:
Xlibris Corporation
1-800-618-969
www.xlibris.com.au
Orders@Xlibris.com.au

Contents

Chapter 1: Our Energy Challenges 4

Chapter 2: Fossil fuels and Challenges 7

Chapter 3: Recycle and Reuse 11

Chapter 4: Renewable Sources of Energy 16

Chapter 5: Realising Opportunities 19

Glossary 22

CHAPTER 1
Our Energy Challenges

Four middle school students Ron, Tim, Liz and Pen are discussing their science project on energy challenges and opportunities.

"What would life be like without energy? Will the whole world be frozen?" observes Tim as he pulls his project file and notebook from his school bag.

"If electricity goes off at home, a lot of things will stop working. There will be no lights, no television, and no computer," responds Pen as she sits next to Tim.

"You are right," agrees Ron. "This is how we can introduce our project. Starting with one big frozen world and investigating how we can unfreeze it is very appealing to me. I am convinced we are the generation that will solve the energy challenges of our time." Ron reaches over and pats Tim's arm.

"Look boys, I think the best place to start this project is to explore and understand what energy is, and highlight some of the energy challenges involved," Liz glances at her best friend Pen with a twinkle in her eyes, as if asking for support.

Pen hesitates, but finally says in a low voice, "Liz has a good idea. A good understanding of what energy is, and then focusing on the energy challenges can be the best way to go."

"I think going over what we have been taught in class and what has been discovered is good practice so we do not reinvent the wheel. But our focus should be thinking outside the box, exploring different ways of understanding these concepts and trying to discover what has been previously overlooked," suggests Tim.

"I agree with Tim, reaching a common understanding first and then focusing on the challenges and opportunities is a terrific idea," concedes Liz.

"So what are the big ideas about energy, the challenges and opportunities,' asks Ron.

Tim and Pen start to peruse through their notebooks. "The sun, for many millions of years before and many millions of years to come will continue to be a major source of energy. The sun gives the earth light and heat energy. Plants and animals need the energy to live and grow. Plants use the energy supplied by the sun to make their own food. Some animals eat some of these plants as food, which provides them with energy," reads Tim.

Pen's eyes brighten as she turns on her notebook. "We cannot see energy. Energy is defined as the capacity to do work and is measured through its effects. The amount of energy is measured using a unit called the Joule. Large amounts of energy are measured in kilo-Joules (kJ). And one kilo-Joule (1 kJ) is equal to one thousand Joules (1000 J)," she summarises.

Liz glances at her notebook and highlights the main points, "There are different forms of energy. Examples are heat energy, light energy, sound energy, kinetic (movement) energy, electric energy, chemical energy, gravitational energy, and nuclear energy. Physical and biological processes interchange these forms of energy. Our bodies need energy to move and grow. Chemical energy in the food we eat gives our bodies energy. When you eat a chocolate bar, you convert the chemical energy in the chocolate bar to useful forms of energy for your body."

"Cars, buses, trains and aeroplanes cannot move without energy from fuels," contributes Pen eagerly sharing a picture of a car at a petrol pump cut from a local magazine. "A lot of energy we use is stored in fuels. A fuel is something that stores chemical energy, like petrol for cars, gas for a cooker/barbeque or wood for a fire. Our mobile phones and other electronic equipment will not work without energy from batteries. Batteries store chemical energy."

"Electricity or electric energy is arguably the most useful form of energy man has learnt to generate. Back in 1831, Michael Faraday showed that when a magnet moves or spins inside a coil of copper wire, an electrical current is generated," contributes Tim.

"The energy needed to move or spin the magnet is turned into electricity," adds Ron, pointing to a diagram on electromagnetism in the textbook.

"Electricity generating power stations use different sources of energy to move or spin the magnet to generate electricity," highlights Liz. "For the dynamo or generator on a bicycle, it is the rider's muscular energy that spins the magnet, generating the electricity that lights the bulb in the headlight."

"The electricity generated in power stations travels cost efficiently through wires to our homes, schools and workplaces to provide lighting and power for other appliances," adds Pen.

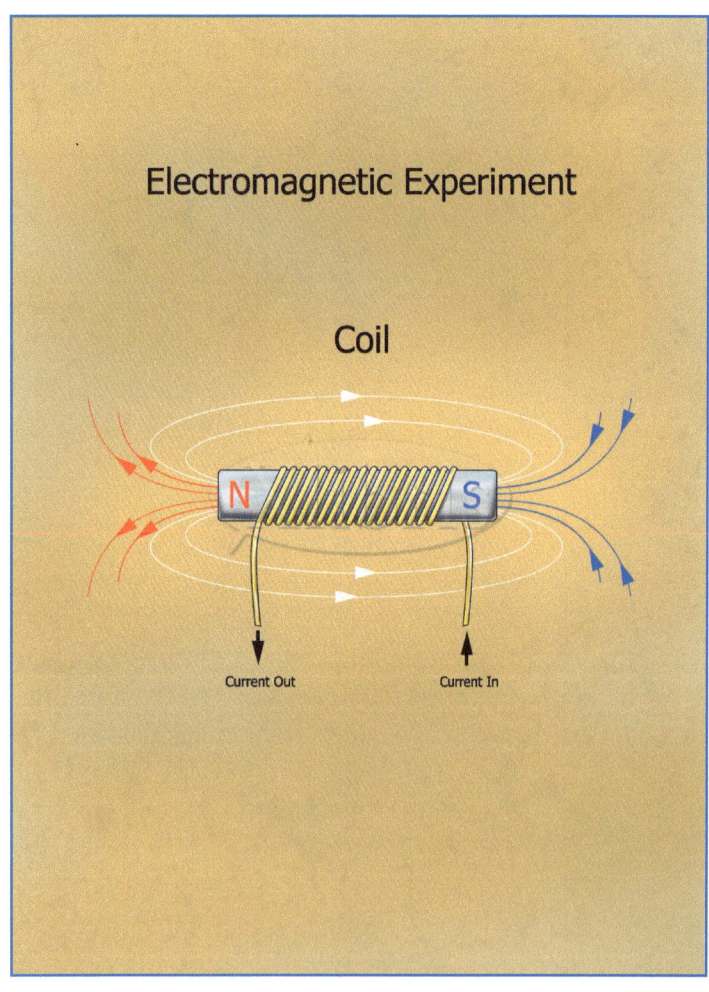

"Some power stations use coal, oil or natural gas to generate electricity," suggests Tim, examining a picture on different sections of a Coal Power Station. "Coal, oil or natural gas is used to heat the water to steam, and the steam spins the magnet. The power stations convert chemical energy in the fuel to kinetic energy and then to electrical energy."

"The most important thing to realise is that we cannot create or destroy energy, but can only change it from one form to another. This is the law of conservation of energy," urges Liz.

"Physical and biological processes basically change energy from one form to another. And we can trace all energy sources to the sun. We need to investigate these energy chains to solve the energy challenges presented to us," adds Ron.

"So what are some of the energy challenges?" asks Tim.

"For a long time we have depended on fossil fuels like coal, oil or natural gas which are non renewable sources of energy. But the world is running out of fossil fuels," argues Pen.

"Oil in particular is running out soon, yet it is the energy source needed to make most of our jet fuels, diesel and petrol, chemicals and plastics. Also when fossil fuels burn, they release gases that some scientists have attributed to causing a number of environmental and healthy issues," urges Liz.

CHAPTER 2
Fossil fuels and Challenges

"So what are fossil fuels and why are we running out of them," asks Tim.

"Fossil fuels are formed under the ground from plants and animals that died millions of years ago. The plants and animals grew and took energy from the sun and stored it. When these organisms died, they were covered by layers of sediment. The pressure on these decayed organisms changed them into fossils fuels which we now use," responds Pen as she peruses her notebook.

"Examples of fossil fuels are coal, oil and natural gas. Remember we cannot create or destroy energy, but change it from one form to another. And once we take the fossil fuels and use them, we cannot replace them. That is why fossil fuels are labelled as non renewable sources of energy because they take a very long time to form and are not reusable," adds Liz slowly. Reflecting on how she and her brother helped their father fire up the barbeque for dinner on the weekend.

"Oil and natural gas are pumped from the seabed and ground. The oil is processed into gasoline or petrol, diesel, kerosene and other products. Coal is dug from the ground. If these processes are not well managed, they can damage the environment," highlights Tim as he points to an image of an offshore oil rig on Pen's notebook.

"A number of oil spills have been reported. One of the worst in the U.S. history is the 2010 BP oil spill in the Gulf of Mexico. These spills have poisoned coastal and wild life. Coal mines have also destroyed large inland areas and disrupted wild life. We have to be very careful about these mining processes," suggests Ron.

"What is Coal Seam Gas (CSG) and why do some people suggest that it is a better substitute for coal," inquires Tim.

"Coal Seam Gas is a natural gas that is extracted from coal seams deep underground," responds Pen as she quickly checks her notebook. "It is predominantly methane, which is an odourless and colourless gas. At the end of the production line, the gas is chilled into liquefied natural gas (LNG) for use as a source of energy and generating electricity."

"Using Coal Seam Gas to generate electricity instead of coal reduces emission of poisonous waste by up to 70%, ensuring cleaner energy supply. But Coal Seam Gas also has its environmental concerns," explains Liz. She refers to her notebook and adds, "Coal Seam Gas mining extracts a lot of water from the ground, and this has a large impact on agricultural activities which depends on this supply of water. Also, the water that is pumped from the ground contains large amounts of chemicals and salts which can harm the environment."

"The main concern is that when fossil fuels burn, they release gases that dirt the air. Carbon dioxide released from burning fossil fuels traps heat in the atmosphere and some scientist have attributed this to global warming, which causes a number of environmental issues. Some heath issues are also linked to the harmful waste products from fossil fuels," highlights Pen.

"The earth is surrounded by a layer of gases called the atmosphere. Some of these gases like carbon dioxide, water vapour, methane, ozone and nitrous oxide are called greenhouse gases because they trap heat from the sun. Excess carbon dioxide can cause the earth to become warmer," expands Liz.

"Carbon dioxide is thought to be the main gas responsible for global warming. Electricity generating plants that burn fossil fuels produce large amounts of carbon dioxide and water vapour. Warm conditions add more water vapour in the atmosphere as heated water from oceans, lakes and rivers turn to water vapour," suggests Tim.

"Methane is released by animals as a waste gas, by rotting plants and rubbish and by burning fossil fuels. Some scientists believe increasing these greenhouse gases from burning fossil fuels can cause climate change and extreme weather. This will adversely affect plants and animals," adds Ron.

"Other that the environmental concerns and health issues associated with our use of non renewable sources of energy; there is the reality that the world is running out of these energy sources," argues Pen.

"Oil and gas have been estimated to run out in 50 years and coal in a few hundred years. Now is the time to think smart about the energy challenges," suggests Tim.

"There is a lot of talk about nuclear fuels. What are nuclear fuels and what challenges do they present?" inquires Ron.

Pen quickly consult her science notes and responds, "Nuclear fuels are pure metals dug from the ground like Uranium. Uranium, like all matter is made of small particles called atoms. But Uranium atoms, unlike most atoms can be split in a process called fission. When these atoms are split in a nuclear reactor, a lot of energy is given out and can be used to generate electricity."

"In nuclear power stations no carbon dioxide or smoke is released. But the problem is that poisonous radio-active waste is produced. This radio-active material is hard to store or dispose and will be harmful to all living things for many hundreds of years. And another concern is the possibility of a nuclear accident," suggests Liz. "For example, the 2011 Fukushima nuclear disaster in Japan that was caused by an earthquake and tsunami," she adds after consulting her notes.

"But some people do not think that global worming or climate change due to burning fossil fuels is a big threat," suggests Ron.

"And others believe that we do not have enough information to make valid conclusions," adds Tim.

"What people should realise is that taking no action cannot be an alternative, considering that the world is running short of fossil fuels and that we are neck deep in our use of fossil fuels," argues Liz.

"We have polluted our earth and atmosphere for many years. It is time we invest in cleaner alternative sources of energy," highlights Pen.

CHAPTER 3

Recycle and Reuse

"Liz suggested that we are neck deep in our use of fossil fuels, why is that," inquires Ron.

"Our addiction to fossil fuels is not only limited to generating energy, but as raw materials for things like detergents, paints, plastics, rubber and many others things," responds Liz. "Most plastics materials we use are made from chemicals processed from oil and natural gas, though others are made from natural materials."

"We use a lot of plastic materials. Some plastics are soft; they can be folded or stretched. They are used to make things like shopping bags or food wraps," adds Tim.

"Other plastics are hard. They are shaped into things like plates, cups, buckets, bins, chairs, tables or toys," suggests Pen.

"We should also realise that most rubber is made in factories using chemicals from oil," highlights Tim. "This man made rubber is called synthetic rubber."

"Natural rubber is made from sticky liquid under the bark of rubber trees. Acid is added to the sticky liquid to extract the rubber. Rubber tree plantations are found in warm countries like Malaysia, Indonesia and Thailand," suggests Liz, as she points to the countries' location on the world map.

"More than 60% of all rubber produced is used to make tyres used in cars, buses, trucks and aeroplanes," suggests Pen.

"Tyres are made from a mixture of rubber from sticky liquid under the bark of rubber trees and rubber from oil," highlights Ron.

"And it is not easy to get rid of old rubber tyres," adds Tim. "If tyre dumps catch fire, they can burn for weeks polluting the environment and poisoning living things."

"Plastics and rubber are useful materials because they do not rot like wood, rust like steel or break like glass," suggests Pen. "But they take millions of years to rot."

"If you throw away pieces of plastic and rubber they end up as rubbish littering the environment for a long time since they do not rot easy," argues Liz.

"The plastics and rubber would be blown into water drains and end up clogging rivers, littering beaches and harming animals and plants," urges Ron.

"We need to realise that oil and natural gas are non renewable sources of energy and will run out in the near future," highlights Pen.

"We should act smart," suggests Tim. "We should use less plastic and rubber materials. We should recycle and reuse plastic and rubber materials instead of throwing them away."

"Almost all plastics and rubber materials can be recycled. The recycled plastics and rubber can be made into other useful materials we use," adds Liz.

"When we recycle we use fewer raw materials," suggests Ron. "And there is less processing which means less greenhouse gases are produced."

"We recycle less than 30% of our plastic materials," highlights Pen after checking from her notebook. "We need to recycle more plastics. Recycling means less damage to our environment."

"We can recycle rubber to produce notebook covers, pencil cases or mats. It is also wise to retread, reuse and recycle old tyres instead of throwing then away," adds Tim pointing to his pencil case.

"We also need to recycle and reuse other materials we use everyday like glass, paper, cloths and metal products," suggests Liz. "A lot of energy is used to collect and process the raw materials to produce these materials."

"Collecting the raw materials involves clearing and digging the earth, and a lot of energy is used. Processing of the raw materials uses a lot of energy and can also damage the environment," highlights Ron.

"For example, glass is made in factories. Pure sand called silica is mixed with limestone (a chalky rock) and soda ash (a chemical)," clarifies Pen after consulting her notebook. "The mixture is heated until it melts and a lot of energy is used. The molten rock is shaped by blowing, pressing or putting into moulds. The molten glass cools and becomes hard and strong."

"Glass has many uses in homes, schools and other places. For example, it can be used as a container for food or drink," suggests Tim.

"We should not throw away used glass bottles or jars," highlights Ron. "Glass cannot be burnt and can take up to a million years to rot. If you put glass in landfills it spoils the environment for many years."

"It is wise to reuse or recycle glass. Empty glass bottles can be returned to bottling plants to be washed, sterilised and refilled," urges Liz.

"Only about 30% of glass is recycled. We need to recycle more glass materials we use," argues Tim. "Reusing glass bottles uses less energy than melting old bottles to make new ones. Melting old bottles to make new ones uses less energy than making new bottles from scratch."

"Most paper we use is made from fibres found in trees like pine or fir. The trees are cut into wood chips, mixed with chemicals and energy is used to heat to pulp. The pulp is taken to a paper mill where it is processed into paper. Energy is used when processing the pulp into paper," reads Pen.

"Paper is used to make books, food containers, wrapping paper, newspapers and many others products," clarifies Ron.

"People throw away newspapers, wrapping paper and food containers after use. And most household rubbish is made of paper products," highlights Tim.

"It takes months for paper to rot in landfills, meanwhile producing methane which is a greenhouse gas that can harm the environment," suggests Liz.

"Paper can be reused in a number of ways to produce artefacts like paper bowls, paper hats and other decorative objects," highlights Ron.

"Used paper can be recycled into new paper. Recycled paper uses fewer chemicals, less water and less energy than making new paper," clarifies Tim.

"Recycling paper also means cutting fewer trees. This helps the environment to become greener," adds Pen.

"We use a lot of metal products," points out Tim. "Examples of metals are iron, copper, zinc, tin, lead, nickel, aluminium and magnesium."

"Metal ores are mined from the ground. The mining processes use energy and can damage the environment. And some of the damage can be very hard to reverse," highlights Liz.

"After the ore is mined from the ground, it is heated until it melts. The ore is refined to extract the metal," clarifies Ron. "The process is called smelting and a lot of energy is used."

"These mining and refining processes produce carbon dioxide which is a greenhouse gas that causes global warming," warns Pen.

"Some metals like aluminium and magnesium are refined using electricity," argues Liz. "The mining and refining of these metals use huge amounts of energy."

"We need to solve our energy challenges and clean up our environment," suggests Tim.

"We have inherited an environment with some pollution and damage caused by industrialisation and other activities," advices Pen.

"We need to clean up the pollution already caused to our environment and make sure we cause minimum pollution going forward," adds Ron. "We should pass this earth to our children and grandchildren with less pollution."

"Recycling metal products saves our environment by using less energy than refining metals from raw materials," proposes Liz.

"We need to reduce, reuse and recycle the materials we use so as to cause less damage to our environment," suggests Pen.

"Reducing means using less of these materials where we can. Reuse means we should use these materials again and again. Recycle means we make something new from the materials," highlights Tim.

"Reducing, reusing and recycling the materials means we use less resources, less energy and produce less greenhouse gases," urges Ron.

"We also need to recover the energy used in the materials that cannot be recycled or reused," suggests Liz. "For example, we can use plastic that cannot be recycled as fuel."

"These processes are not new, but they imply change in behaviour from everyone, which is not an easy thing," urges Ron.

"The processes of reducing, reusing and recycling the materials, and recovering the energy from materials that cannot be recycled or reused can only be effective if we all get involved,' highlights Pen.

"We have a responsibility to one another as a human race," adds Tim. "Everyone has to get involved."

CHAPTER 4
Renewable Sources of Energy

"Renewable sources of energy do not get used up. We can use these resources again and again and again," clarifies Liz. "These facts mean that the need to develop renewable sources of energy is incontestable."

"Renewable sources of energy are labelled as clean or green sources of energy because they release little or no greenhouse gases to produce electricity," summarises Pen from her notes. "And with world wide concerns on global warming and climate change largely due to burning fossil fuels, renewable sources of energy need to play a major part as energy sources," she adds.

"Examples of renewable sources of energy are solar (sun) energy, wind energy, water (hydro) energy, and energy from plants and animals," clarifies Ron.

"However, renewable sources of energy are not without challenges," suggests Liz. "They depend on unpredictable naturally occurring weather events such as wind and sunshine, hence the need for cost efficient methods and technologies when harnessing and storing the energy from these sources."

"A number of different storage battery materials and technologies have been identified to store the energy from solar, wind or water waves so we can use that energy at times when the sun doesn't shine or wind doesn't blow," highlights Tim.

"But the challenge is to modify these materials and technologies to achieve long term stability and cost efficiency," adds Pen.

"The energy from the sun has been around for many millions of years and will be around for many millions of years to come," suggests Tim. "This solar energy is found everywhere on the earth's surface but is currently more expensive to harness than non renewable energy."

"We use solar energy to heat water and generate electricity," adds Liz. "Solar panels are black metal plates attached to water pipes and used to heat the water. Solar cells convert light energy to electric energy."

"Solar cells are mostly made from silicon, which is a common substance extracted from sand. Unlike in generators, no spinning of the magnet is involved. But at present solar cells are very expensive to make," argues Pen.

"Solar cells are used to supply electricity in homes. Small solar cells are used to supply energy to watches and calculators. Solar cells are also called photo-voltaic (PV) cells," highlights Ron.

"Wind power has been used for many years. Wind pumps have been used to pump water from the ground. Wind power has also been used to sail boats," points out Ron.

"All winds are caused by the sun," suggests Tim. "Wind energy can be used to spin turbines. Turbines are placed in windy areas and have blades that are turned by the power of winds. The blades are linked to generators which produce electricity."

"If the blades are longer, they generate more electricity. If the blades turn faster, they generate more electricity," adds Pen.

"Groups of wind turbines are called wind farms. Some wind farms have been placed short distances from coastal areas. These wind farms can be noisy and can disturb local environments," highlights Liz.

"When water energy is used to generate electricity, it is known as hydro-electric energy. Energy from falling water, water currents, water waves and hot water from the ground are renewable sources of energy," suggests Tim.

"Water falling over a waterfalls or dam wall can be used to turn turbines which generate electricity. Water turbines can also be turned by energy from flowing water or water waves to generate electricity," explains Pen pointing to an image of a hydro-electric power station.

"In some parts of the world, hot rocks deep in the ground can heat water in the ground to create steam. The steam can be used to turn the turbines and generate electricity. This energy is called geothermal energy," adds Ron.

"Bio-fuels come from living things like dried leaves, dung from animals or oils removed from plants like sunflower or nuts," suggests Liz. "Traditionally people used dung and fat from animals as fuel."

"Fuels extracted from plants are renewable sources of energy because new plants can be planted. Some of these plants are sugar cane, soya beans and wood. These plant fuels depend on energy from the sun," adds Tim.

"The problem that can arise with extracting bio-fuels is that some plants are used for food, or might take up farming land and can result in food shortages in the future," observes Pen.

"We can also harness energy from waste products. Methane gas is released from manure, crop waste and rotting rubbish and can be used as a source of energy," suggests Ron.

"A very small amount of the world's power comes from renewable sources of energy, because many countries are still depended on fossil fuels," suggests Liz.

"Many people think it is too difficult and expensive to switch to renewable energy," highlights Ron.

"We really need to think smart about the energy challenges, especially considering that economic and political factors come into play when we consider alternative energy sources," highlights Pen.

CHAPTER 5
Realising Opportunities

"Bill Gates, the founder of Microsoft suggests that if we don't have innovations in energy we don't improve much at all. Why would he say that?" asks Ron.

"The world's population is growing, so more energy is going to be needed. But we are going to run out of coal, oil and gas in the near future," replies Tim. "This is why we must find alternative sources of energy that cause less damage to the environment. Continuing to do the same cannot be an option."

"It is essential to find innovative ways to generate energy without polluting the environment," points out Liz. "We have to make better use of renewable sources of energy and improve the efficiency of existing power stations that use non renewable sources of energy."

"For example, when power stations burn coal, only about 30% of the energy is turned into electricity and the rest is wasted as heat, though heat exchangers in some power stations use some of this energy to heat water," summarises Pen from her notebook.

"We need to innovate more efficient power station and energy efficient buildings: homes, schools and workplaces. Home owners can save energy and money by insulating their homes," adds Tim

"We need to invent more commercial and industrial cost effective and efficient technologies. For example solar cells are expensive to make, and only about 10% of the solar energy is converted to electricity," argues Ron.

"Big cars, old cars and faulty cars use large amounts of fuel. These can be replaced by Hybrid cars that use less fuel and reduce harmful gas emissions," suggests Liz.

"Light-Emitting Diode (LED) lights and LED screens use less electricity than light bulbs and TV screens," argues Tim. "These can be used to replace light bulbs and TV screens because almost all the energy used in LED technology is turned into light."

"Battery cells can be replaced by fuel cells that use hydrogen gas if we can make the technology cheaper and smaller. Batteries take a lot of energy to make and create harmful waste compared to hydrogen cells," highlights Pen.

"Improving technology should not be the only solution. We need to educate the public and change how people use energy. Bearing in mind that changing the way people behave is not easy," suggests Tim.

"We all need to think and act smarter. Everyone can reduce the amount of energy they use by changing their behaviour, for example cycle rather than drive, insulate homes so heat does not escape, turn off switches, wear warm cloths rather than turn up heating, recycle and use gargets with less energy consumptions," highlights Ron.

"This will reduce the need to build more energy wasting and polluting power station as we move to new energy sources of the future," points out Pen. "As long as light and heat energy from the sun reaches the earth, renewable sources of energy will never run out."

"A gradual move to cost effective, efficient and non polluting energy sources can be the way to go for our project on energy possibilities for the future," suggests Liz.

"This can include solar energy, wind energy, water (hydro) energy and hydrogen fuel. These resources are abundant, but the technologies to harness and store energy from them is yet to be cost effective and efficient," adds Pen. "We should explore some of these possibilities for our project:

1. Explore cost effective, efficient and socially and environmentally friendly ways to generate and store solar energy.

2. Explore cost effective, efficient and socially and environmentally friendly ways to generate and store wind energy.

3. Explore cost effective, efficient and socially and environmentally friendly ways to generate and store water (hydro) energy.

4. Explore cost effective and efficient ways to separate hydrogen gas from water. Water is made of hydrogen and oxygen gases. When hydrogen gas burns in air, it produces large amounts of energy plus water."

Get into a team with two or three friends. Choose one or more from the four energy possibilities of the future, or any other energy related challenge of our times. Explore cost effective, efficient and socially and environmentally friendly ways to harness and store energy from the source(s). You are part of the generation that is going to solve the energy challenges of our times and explore possibilities to make this world a better place.

Glossary

Battery – cells that stores chemical energy

Bio-fuels – fuels extracted from living things, plants and animal

Coal Seam Gas (CSG) – a natural gas extracted from coal seams in the ground

Dynamo – changes kinetic energy to electrical energy

Energy – capacity to do work or make things happen

Fossil fuels – sources of energy formed millions of years ago from dead plants and animals

Generator – changes kinetic energy to electrical energy

Hybrid car – uses electric motor at low speeds and petrol engine at high speeds

Joule – unit used to measure energy

Light-Emitting Diodes (LEDs) technology – less energy is wasted

Non renewable – something that gets used up

Organisms – living things, plants and animals

Pollution – harmful materials that contaminate the environment

Renewable – something that does not get used up

Solar cell – converts light energy to electric energy

Solar panel – converts light energy to heat energy

Turbine – machine with blades which are turned by wind, water, steam, etc, to produce electricity